AMAZING
SCHEMES
WITHIN YOUR
GENES

Written by
Dr Fran Balkwill
illustrated by
Mic Rolph

Collins

An Imprint of HarperCollins*Publishers*

First published 1993
Reprint 1994
© text Fran Balkwill 1993
© illustrations Mic Rolph 1993

A CIP catalogue record of this book is available from the British Library
ISBN 0 00 196465-8 (PB)
0 00 196466-6 (HB)
Thanks are due to the Science Photo Library for permission to use the illustrations on pages 24-25

Printed and bound in Hong Kong
This book is set in Lubalin Graph 13/16

Think of all the people you know and see every day. Your family, your friends, people who live in your street or work near you – even people on the radio and television. Then think of all the people who live in villages, towns and cities throughout your country. Try to imagine all the millions and millions of people who live in cities, towns and villages around the world. There are more than five billion other children, men and women, sharing the planet with you. If you could line them all up, the queue would stretch from the earth to the moon – and back again – six times! And the amazing fact is....

...not one of those **five billion** other people looks, thinks or behaves exactly like you!
You are completely unique.

4

Well, to make you the way you are, there are some amazing schemes within your genes.

So what are genes?

5

Genes are recipes for making proteins. Proteins are very important substances in your body. Proteins make the cells in your body the shape and the colour they are. Proteins help cells do all the complicated jobs they have to do. You have about fifty thousand genes inside you. Genes for the many different proteins that make your eyes, ears, nose and mouth; genes for proteins that make your finger and toe nails, teeth and hair; genes for proteins that make those complicated parts like your brain, lungs and heart; genes for proteins that influence your height, your sense of taste and smell, even genes for the proteins that decide your skin and hair colour.

Where exactly do you find genes?

Errr!...

Your genes are in each of the hundred million, million cells in your body.

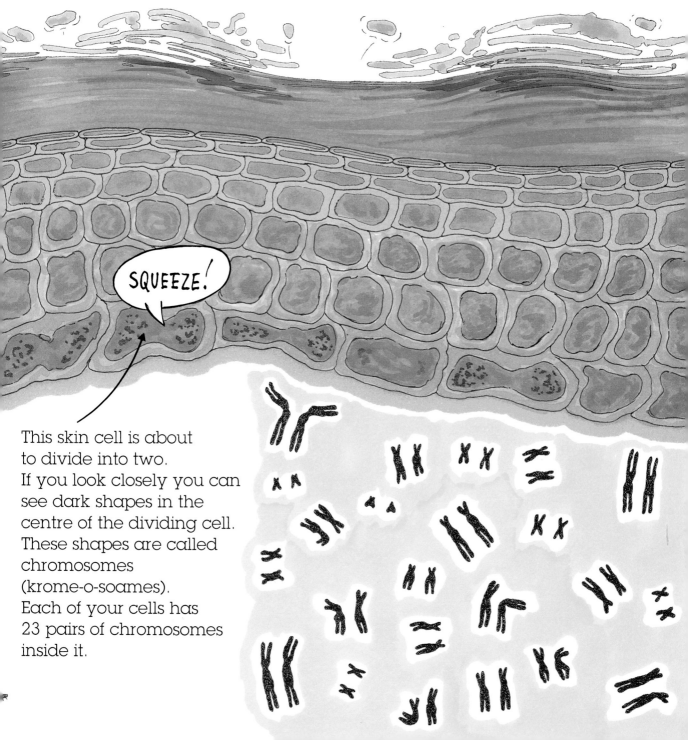

SQUEEZE!

This skin cell is about to divide into two. If you look closely you can see dark shapes in the centre of the dividing cell. These shapes are called chromosomes (krome-o-soames). Each of your cells has 23 pairs of chromosomes inside it.

If you *could* unravel one chromosome from one of your cells and look at it under a mega-powerful electron microscope, you would find that chromosomes are made of a thread-like substance called deoxyribonucleic acid (dee-oxy-rybo-new-clay-ic) acid – everyone calls it DNA for short.

Genes are made of this DNA.

If each gene is a recipe, then DNA is a chemical language that the recipes are written in. There are gene recipes all along the DNA threads that are wound up in each chromosome. Gene recipes are made from just four different chemicals joined to each other along the DNA strand. They are called –

Adenine (add-en-een),
Thymine (thy-meen),
Cytosine (cy-toe-seen),
Guanine (gwa-neen).

Scientists call them **A T C** and **G**. Each one is drawn in a different colour.

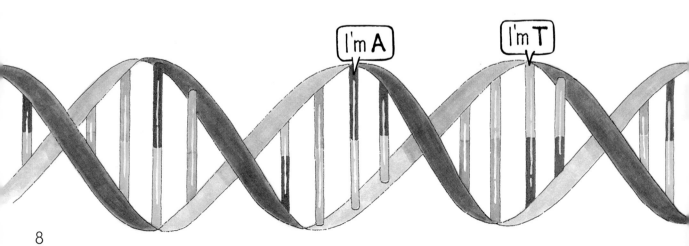

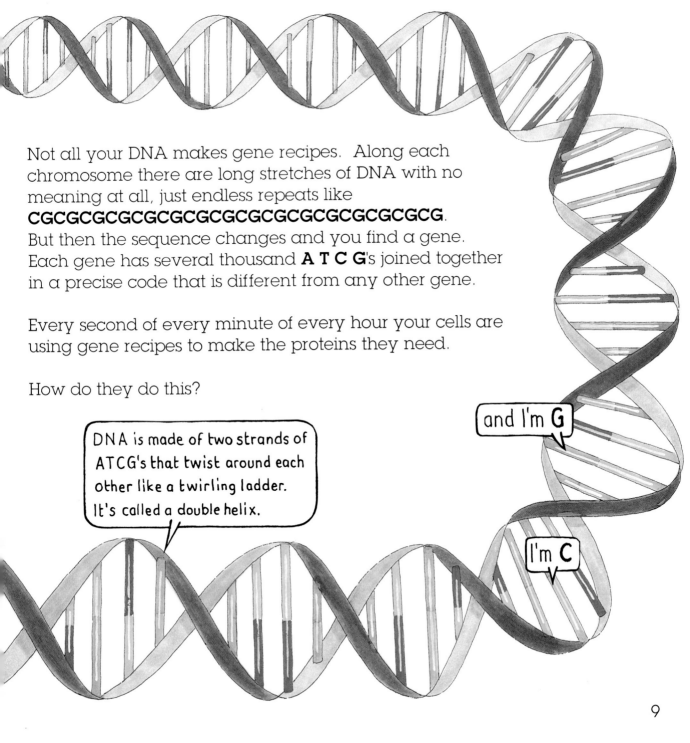

Not all your DNA makes gene recipes. Along each chromosome there are long stretches of DNA with no meaning at all, just endless repeats like **CGCGCGCGCGCGCGCGCGCGCGCGCGCG**. But then the sequence changes and you find a gene. Each gene has several thousand **A T C G**'s joined together in a precise code that is different from any other gene.

Every second of every minute of every hour your cells are using gene recipes to make the proteins they need.

How do they do this?

and I'm **G**

DNA is made of two strands of ATCG's that twist around each other like a twirling ladder. It's called a double helix.

I'm **C**

1 When a protein is needed, the part of the DNA that is the gene for that protein unwinds from its particular chromosome.

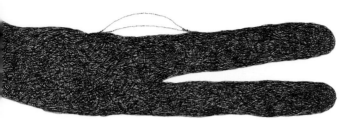

2 Then a copy of the gene is made from one of the two DNA strands.

3 The copy strand floats off to another part of the cell. Here it acts as a pattern to build a protein.

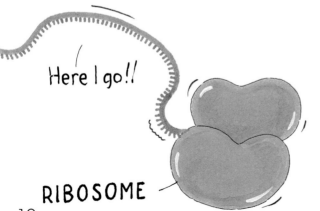

Here I go!!

RIBOSOME

4 Proteins are assembled from building blocks called amino acids that float around inside the cell.

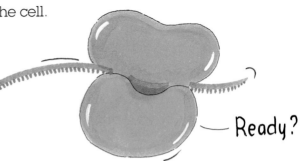

Ready?

5 The copy strand tells the cell to join up the amino acids in the correct order, like the beads on a very long necklace. The necklace of amino acids then folds up tightly in a shape that is different for each protein.

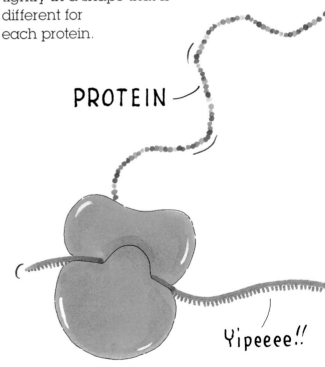

PROTEIN

Y'ipeeee!!

So, as you read this book remember that:

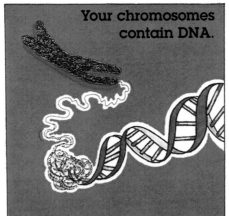

Your chromosomes contain DNA.

SNORE!!

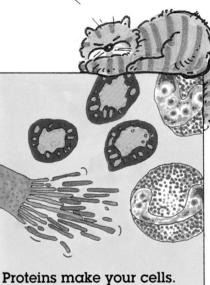

DNA makes genes.

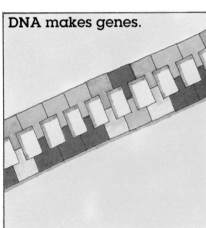

Genes are recipes for proteins.

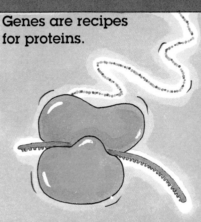

Proteins make your cells.

Your cells make you.

And the reason why you are different from everybody else is because your mixture of genes is slightly different from their mixture.

11

How did you get your unique mixture of genes? You were created when two cells fused together – one cell from your father (the sperm), one cell from your mother (the egg). Sperm and eggs are made by a very special type of cell division. They don't have 23 pairs of chromosomes like all the other cells in your body, they each have only 23 single chromosomes.

This is the really clever bit!

When the sperm or eggs are being made, each pair of chromosomes interlocks and genes are swapped from one chromosome to the other. This means that each individual egg or sperm contains a mixture of genes that is different from every other egg or sperm – in fact it is completely original.

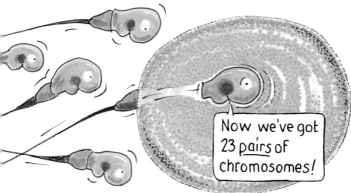

Now we've got 23 pairs of chromosomes!

When the sperm and the egg join together, the first cell of the new human being has 23 pairs of chromosomes again, 23 from the egg and 23 from the sperm. So your 46 chromosomes are a mixture of your parents' chromosomes, and your genes are a mixture of their genes. But your mixture of genes is different from both of your parents. Why?

I didn't know that!

12

This how it works

Let's look at just two pairs of chromosomes – coloured red and blue.

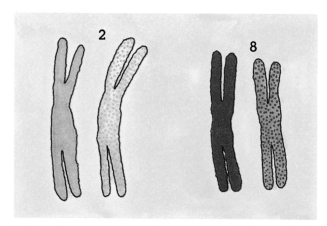

1.The chromosomes have copied their DNA and now get shorter and thicker. They get very close to each other.

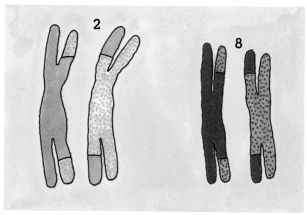

3.When the chromosome pairs finally part, they have exchanged some of their DNA. They each have a different mixture of genes from the original pair.

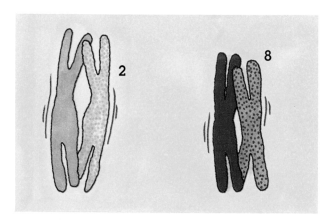

2.The arms of the chromosomes coil round each other and can stay like this for a very long time. All eggs in a girl reach this stage before she is born, and the arms remain coiled around each other until the child matures, ten or fifteen years later.

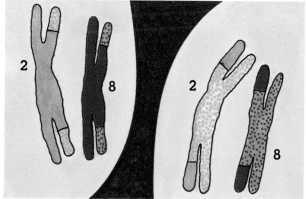

4.But the egg and sperm cells still have 23 pairs of chromosomes. What happens next? Well, the egg cell divides again, but this time one chromosome from each pair goes into each cell. **The egg and sperm are finally ready to make a new human being!**

He's about to find out the difference between boys and girls!!

When you started your life, you were made of just one tiny cell, with 23 pairs of chromosomes containing a unique mixture of genes. But now that same mixture of genes is in the millions and millions of cells that make up your body. How does this happen? Well, you grew because that first cell became two cells, those two cells became four, four became eight and so on until you were made of millions of cells. Before each cell divided, the DNA that makes your genes was copied, so that each new cell would contain your genes as well.

CRINGE SWEAT/ GULP.

How do genes make the difference between this girl and boy?

Just one gene on one special chromosome makes the difference! If we take a cell from each child and look down a microscope at their chromosomes, we can sort out the 23 pairs – in the girl. But we can only make 22 pairs from the boy. What about the other pair? The chromosomes don't look alike! One is shaped like an **X** and the other is smaller and like an upside down **Y**.

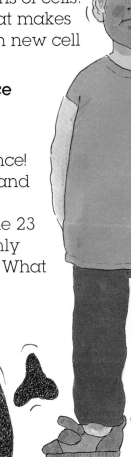

life line

protect the rainforests

All human embryos start developing in the same way. If the embryo has two **X** chromosomes, an area of cells becomes the egg-making part (the ovaries). The baby will be a girl. If the embryo has an **X** and a **Y** chromosome then a protein made from one gene on the **Y** chromosome, signals some of the cells to start forming the sperm-making part (the testes). Once this has happened, the baby will be a boy. Many more genes then make many more proteins that cause the differences that you can see between girls and boys, but without that one gene on the **Y** chromosome – we would all be girls.

Just imagine.

Of course, human beings are different from each other in many many ways. It is not always easy to understand the influence of genes on these differences but there are some very obvious clues. Have a good look at all the families you know and see if you can find some of these inherited features. Now there is no need to be bored at a large family gathering like a wedding or a birthday party! Play the genetic detective and see if you can spot other inherited differences, and guess who is related to who.

Occasionally you can find very strong evidence for one particular pattern of inheritance. (Clue: look for a family in which both parents have red hair). But most inheritance is not so simple. Most differences between human beings are controlled by more than one gene – height and skin colour, for example. Another reason for the complicated patterns of inheritance is that we all have **two** copies of every gene on each pair of chromosomes, one from our mother and one from our father. Both copies can be used by your cells.

And
you must remember that the way you live, the food you eat, all the many experiences you have, act upon the raw material of your genes to make you as you are.

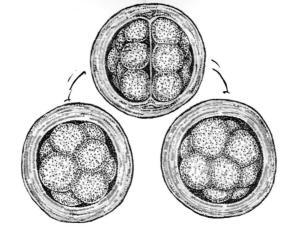

Is it really true that every human being is unique? What about twins? Do you know any? Maybe you are one! There are two types of twins, identical and non-identical. Identical twins have identical genes. During the first two weeks of life the tiny ball of cells that is a new human being, occasionally splits into two. From that point on, two babies begin to develop – after they are born even their mother can't tell them apart!

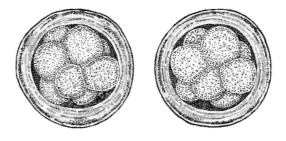

Identical twins are the same sex, they have the same colour eyes and hair, and the shape of their noses and ears is identical. Even if identical twins are separated at birth and raised in different families they can develop in remarkably similar ways.

But not all twins have identical genes. Non-identical twins happen when the mother releases two eggs at the same time. So two sperm fuse with two eggs to make two babies with quite different mixtures of genes. These twins are no more like each other than brothers and sisters except that they share a birthday.

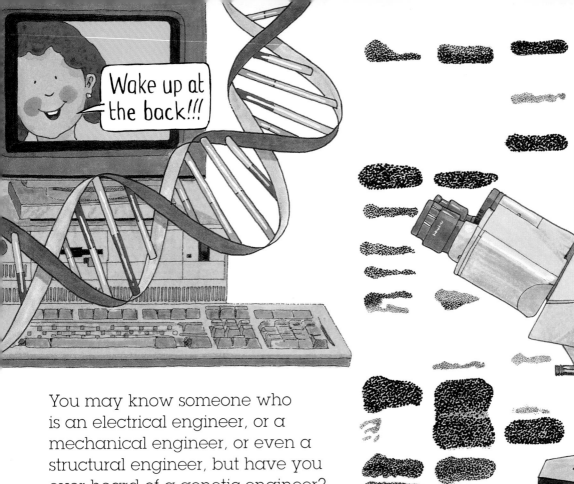

You may know someone who is an electrical engineer, or a mechanical engineer, or even a structural engineer, but have you ever heard of a genetic engineer? There are many thousands in our world today and the experiments they are doing now will undoubtedly affect your life as you grow up. Can you believe that genetic engineers can take human genes – each about one millionth of a metre long and one thousand millionth of a metre wide – and put them into bacteria?

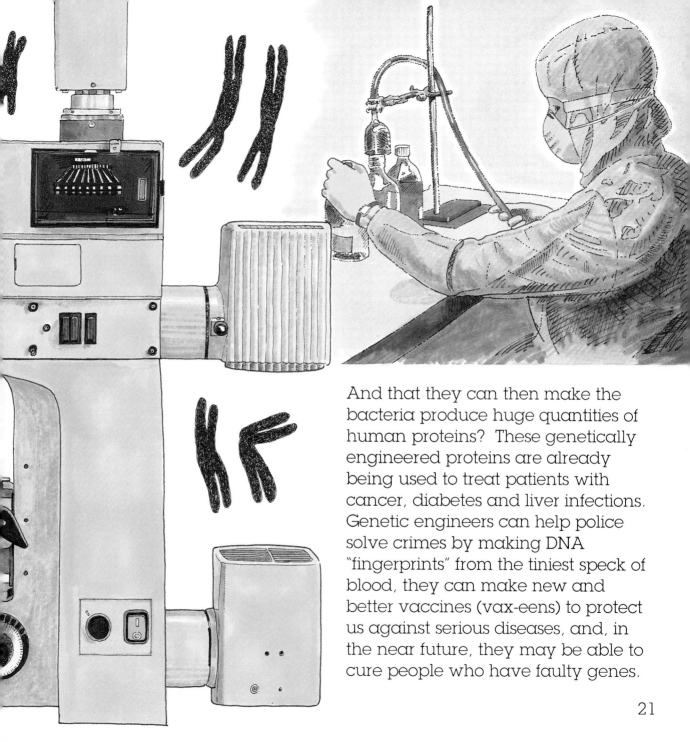

And that they can then make the bacteria produce huge quantities of human proteins? These genetically engineered proteins are already being used to treat patients with cancer, diabetes and liver infections. Genetic engineers can help police solve crimes by making DNA "fingerprints" from the tiniest speck of blood, they can make new and better vaccines (vax-eens) to protect us against serious diseases, and, in the near future, they may be able to cure people who have faulty genes.

We've already told you that every time one of your cells divides into two, a copy is made of all your genes. A faulty gene happens when a mistake is made by a cell as it copies a gene or when DNA is damaged. Mistakes in genes are called mutations (mew-tay-shones) – and they are happening to you all the time. Cells with a faulty gene may not do their jobs properly, but you have one hundred million, million cells in your body. Most cells with a faulty gene will soon die and be replaced by healthy ones.

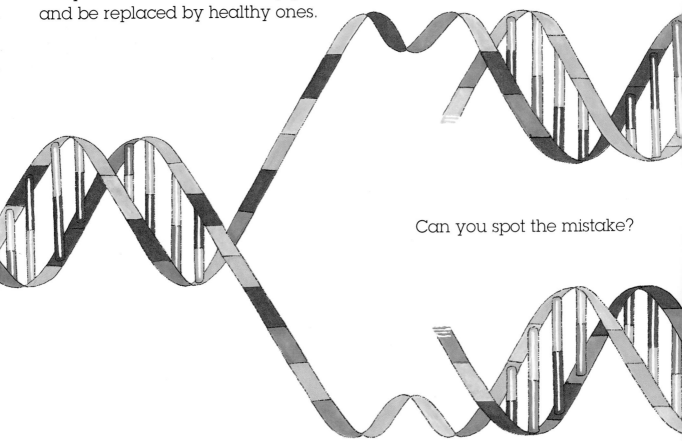

Can you spot the mistake?

But what happens if there are mutations in the genes of sperm or egg cells? If sperm or egg cells with a faulty gene do not die, they may make new human beings. In this way, mutations can be passed on to the next generation. Some inherited mutations don't cause much trouble. For instance, you may know someone who is colour blind...

Can you see the shapes inside the circles? If you can't, you have probably inherited colour blindness. Most people can see three basic colours – red, blue, and green. This allows them to see all the different colours of our world. But some people have an inherited mutation in one of the genes for an eye cell protein. Their eye cells don't work in the same way. Most children who have colour blindness cannot tell the difference between red and green – (embarrassing if you support Liverpool or Celtic!). A few children confuse blue and yellow. Very rarely some children are totally colour blind, and their world is nothing but shades of grey.

Another job for the genetic detective – do our colour test on all your friends, classmates, and their families!

If you test enough children you may notice that most children with colour blindness are boys.
Do you know why?

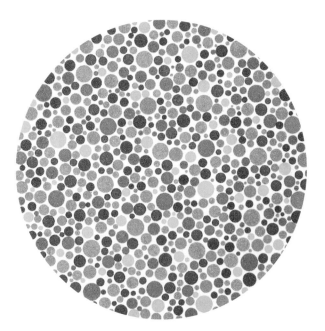

Clue: The faulty gene is on the **X** chromosome. You may remember that the **Y** chromosome is smaller than the **X**. Because of this, boys have only one copy of many of the genes found on the **X** chromosome. Now do you realise why boys are more likely to be colour blind?

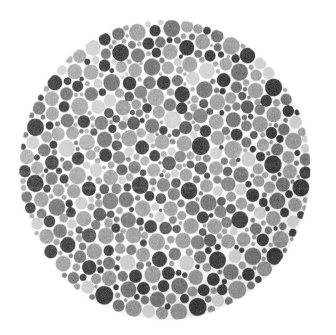

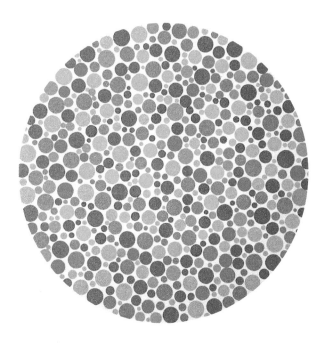

But other inherited mutations can make people ill. There are over five thousand different diseases in the human race that are caused by inherited mutations in our genes.

One of these illnesses is called cystic fibrosis (sis-tick fye-bro-sis). Children with this have one faulty gene. This gene makes a protein that normally helps the lining cells of the lungs and digestive system keep the tubes clear, moist, and full of sticky mucus to trap germs. Because this protein doesn't work properly, the lungs of children with cystic fibrosis become full of thick mucus. Bacteria grow there and make the children ill. Part of their digestive system gets blocked and they can't digest their food properly. Nowadays nurses and doctors have clever treatments to keep children with cystic fibrosis healthy and help them lead as normal a life as possible. And in the future, genetic engineers may be able to replace the faulty gene and make the children better.

Scientists have discovered that the faulty gene for cystic fibrosis is found on chromosome 7.

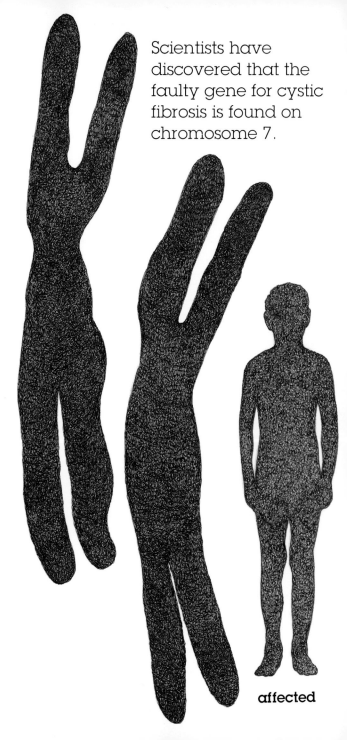

affected

How do children get cystic fibrosis? In Britain, Northern Europe and America, about one person in twenty-five has one faulty gene and one normal gene for this protein. They are perfectly healthy because the one correct gene makes all the correct proteins their cells need. But if a mother and father both have one faulty gene, some of their children could inherit two faulty genes and those children will be born with cystic fibrosis.

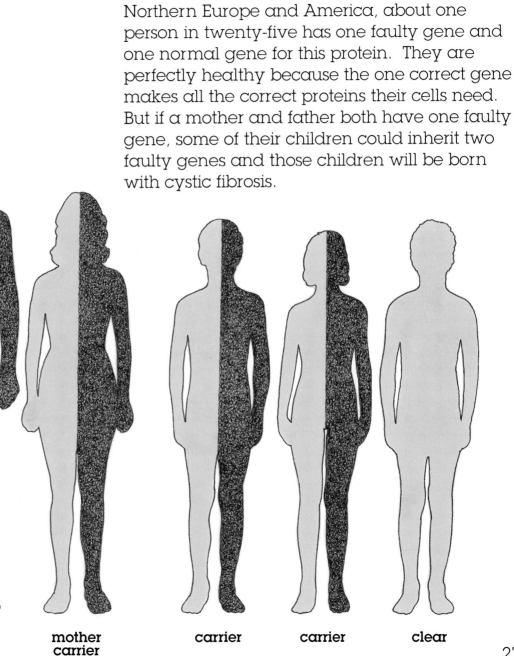

father carrier   **mother carrier**   **carrier**   **carrier**   **clear**

27

Millions of people in the world suffer from sickle cell anaemia (an-ee-me-a). This illness is caused by a mutation in the gene for haemoglobin (heem-o-glow-bin). Haemoglobin is the protein in red blood cells that carries oxygen from your lungs to the rest of your body. Red blood cells are normally round and smooth, but some red cells from children with sickle cell anaemia are long, curved and thin because the faulty haemoglobin does not keep them the right shape. These are called sickle cells because of their shape. Sickle cells block up tiny blood vessels and don't live as long as normal red cells. This means that children with the disease often get tired and ill, and are in pain. Nowadays doctors and nurses have many ways to help these children.

Children with sickle cell anaemia have two copies of the faulty gene – one inherited from each of their parents. One in ten people who live in, or have ancestors who come from, Africa or the Caribbean, have one normal and one faulty gene for haemoglobin. This disease is also common among people from the Middle East, the Mediterranean, and Asia.

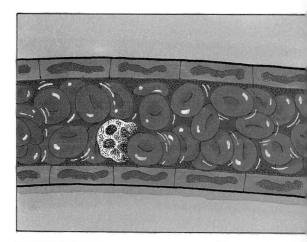

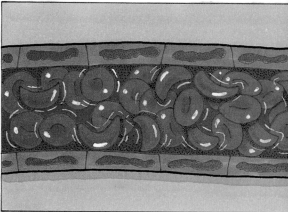

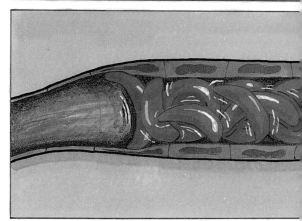

How did sickle cell anaemia start?

About ten thousand years ago when some early human beings lived in hot and humid regions of the earth, a deadly disease arose. We now call that disease malaria (mal-air-i-a). The malaria germ grew in the liver and red blood cells of humans. It was easily spread by mosquitoes who bit infected people. Malaria made people feverish and so weak that many of them died. But after several hundred years there were a few people who didn't get malaria, or didn't get it so badly. This was because a mutation had happened in one of their haemoglobin genes. Although people with one normal and one faulty gene for haemoglobin were perfectly healthy, their red cells were slightly altered and the malaria germ could not easily grow in them. Thousands of years passed. As you might expect, people with one faulty haemoglobin gene lived longer than others because they were resistant to malaria. So these people had more children, who also inherited the faulty gene. Soon children were born to parents who both had one faulty gene. Some of their children would have two copies of the faulty gene and be born with sickle cell anaemia.

GLUG!!

Although mutations can be bad, they can be useful as well. All forms of life on this planet have genes made of DNA in their cells. Mutations allow all living creatures to change and adapt to new surroundings. Mutations create new species. Mutations are the reason why life on earth has evolved from a single-celled ancestor into millions of different plants and animals. Take you, for instance. You are a member of the species homo sapiens. Your closest living relatives are the apes. In fact 98.4% of your genes are the same as the genes of any chimpanzee you might see in a wildlife park.

About twenty million years ago your ancestors lived in tropical woodland. Their bodies were well adapted to climbing trees, and they lived and hunted in social groups. Three or four million years ago useful mutations happened so that some of your ancestors were able to walk upright on two legs. The shape of the bones in their hips, knees and feet changed and their arms became shorter than their legs. Then they could travel further to find food and move faster to escape hungry predators. Their hands, freed from helping them move along the ground, developed a powerful, precise grip. Between one and two million years ago, more useful mutations gave these apes larger brains.

Then they invented many tools from the wood and stone around them. They began to communicate and help each other hunt for food.

Just 40,000 years ago, the first true members of our species began to populate the planet.

Scientists studying our genes now know that we are all descended from that small group of early humans. Whether our skins are light or dark, our hair black or fair, whether we now live in the Arctic or at the equator, the message of our genes tells us that we are brothers and sisters...

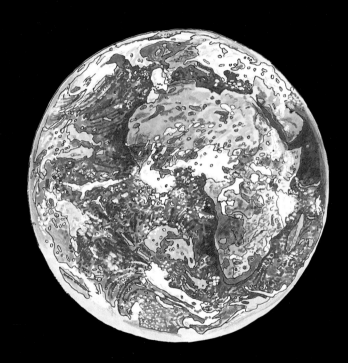

all five billion of us!